Origin of Black Hole
Beginning of universe

ABOUT THE AUTHOR

Akshay Kumar is an author , tutor, teacher and inventor. Inventor of chromatography machine with the team of scientist and professor and patent accepted and published by govt. of India. And studied it from prestigious institution ISRO and NASA and completed its professional degree and diploma courses. Attained number of webinar and conferences on national and international level which helped a-lot in his research. Currently working as a teacher in green field public school sihunta chamba Himachal Pradesh India.

ACKNOWLEDGEMENT OR DEDICATION

In the Act of doing this work successfully, I am very thankful to those persons who encouraging me to do this. Firstly I am thankful to God for Deemed me worthy .By the god grace this work is done. Then thanks to Dr. Surinder Paul who motivate me at all times , and encourage me in research work , He assist me all time . So I am sincere gratitude to my research supervisor , Dr. surinder Paul (Associate Professor) at Central university of Himachal Pradesh. Special thanks to Dr. Dinesh Sharma (Officiating principal Govt. Arya degree college Nurpur, (H. P), Dr. Dinesh Pathak (Professor at university of west indies. ,Dr.Rohit Sharma, (Assistant Professor)

Department of Applied Science

Satyam Institute of Engineering & Technology, Amritsar ,he is that person who help me a-lot at all times in every area in any cost. Because without the blessings of teachers this work is not be done.

Preface

I feel immense pleasure in presenting this book "The origin of black hole" for all students, teachers , researchers, and scientists.

The text has been written in a simple and lucid language in a brief manner. All the conclusion is based on previous research in this field.

This book is basically depends upon different theory like theory of relativity and current research.

I hope that the book will be able to fulfill the objectives for which it has been written. Suggestions for further improvement of this book will be highly acknowledged.

Special thanks to my all professor and teachers.

AKSHAY KUMAR

ABSTRACT

A Black Hole is zone of space time where gravitational force (G) or force of gravity is very strong that neither can be elopement from it , Even light of electromagnetic Radiation cannot be escape from black hole . Black hole is a terrible truth about the universe. Black hole definition expressed that it's black region inside the galaxies . The Word Black Hole firstly Coined by John Wheeler a American theoretical Physicist in 1969 . A Black Hole is formed when a star Is collapse due to its own gravity (g) or inter-atomic (IA) repulsion of matter .When a star is contract due to its own gravity (g) ,whole of the mass of a deadly star is imposed in a point , called the Singularity point . So resultant ,this point quit unstable and that is why according to " Pauli Exclusions principle the matter having the same atomic number cannot stay in a same atomic level ". So here Repulsion occur in a matter, and this cause the explosion is known as " Supernova " . Now after the supernova explosion the gravitational force and exclusion repulsion is in its state of equivalent for a few times . But after some time this gravitational force is comparably large as compared to exclusion repulsion . so due to high gravity the length of a star contract and occur in the form of Singularity . One question Always arise in our mind that a light which is Electromagnetic waves is in the form of " Photon" which is mass less then why light can be attracted by a black hole , in simply manner then why light cannot be escape from Black hole ?Because in Newton law of universal gravitational constant "Newton states that - gravitational force is always relative to the masses of the body and inversely relative to the square of the distance between them . So, answer of this question is comes from " Einstein General Theory of Relativity" , which states that light cannot be attracted by Black hole ,but Black Hole divert the path of the light and due to the curve in space time the path of the light will be bend at its edge of the black hole and this edge of the black hole is known as Event Horizon .And due to this twist in space time ,light falls down in the region of Black hole . So black hole is an peculiar object in our universe . There is a lots of Black hole present in our universe . Also in our milky way galaxy , one black Hole is present which is known as " Sagittarius A* " . This Black Hole Present at the Galactic center of the Milky way Galaxy . So I simply called Black hole as a Ghost where they can eat everything which comes around . Before to know About the Black Hole it is necessary to know about our Universe , How our universe came into Existence ? How a star Collapse and form a Black Hole ? we can solve these problem only when we know about the History of our Universe from " Big Bang to Black Hole ". so this is short inception about Black Hole, we discuss Black hole further.

Singularity point and Event Horizon

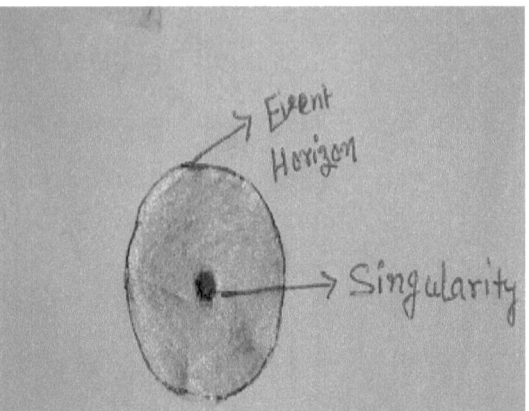

Contents

Edwin Hubble (1889-1953)

BIG BANG THEORY

BIG BANG IMAGE

CHAPTER - 3

BLACK HOLE

John Michell an Astronomer and Natural Philosopher (1724 – 1793)

Subrahmanyan Chandrasekhar an Indian – American Astrophysicist (1910 – 1995)

A white dwarf Sirius B which is orbiting around Sirius A.

Robert Oppenheimer (1904 – 1967)

3.1 A SHORT LIST OF KNOWN BLACK HOLES

3.2 The Nearest Stellar Black Hole

The Nearest Stellar Black Hole

3.3 The Nearest Super- massive Black Holes

The Nearest Super-massive Black Hole

3.4 Moon as a Black Hole: -

3.5 The Milky Way as a Black Hole:-

3.6 The Event Horizon and Singularity Point : -

Singularity Point : -

Event Horizon : -

Interior Space : -

Event Horizon

Singularity point

3.7   Gas Temperature of Black Hole

Black Hole as emitting Ray of radiation or light

3.8   Time Dilation near the Earth and Black Hole: -

3.9   Accretion Disk Temperature Of a Black Hole : -

Black Hole Accretion Disk with Radiate Heat

4.0 Distortion near the Black Hole: -

Distortion near the Black Hole

4.1 Gravitational Field near the Black Hole: -

This curve shows that gravity will be effect on space time.

Black hole with its high Gravitational field

4.2 Tidal Force on the surface of Black Hole: -

Tidal Force on the surface of Black Hole

4.3 A star Collapsing into a Black Hole : -

Star Collapsing into a Black Hole

INTRODUCTION

Black hole is a region in which gravitational force is very strong so , that neither can be escape from it . Even A light having velocity three million kilometer per second cannot be escape from a black hole. So I call A black hole simply a Ghost who cannot left anything around it. Before the Albert Einstein ,space and time is a mystery. Because at that time Newton law of gravitational is used to define the motion of a heavenly Body . At that time Newton Theory is applicable which describe that Space And Time is Absolute, which means that Time cannot change anywhere in a space .it is same in everywhere in a space. But when Albert Einstein things about that concept of space and time , He Realize that space and time were not absolute , it is Relative . He used an Experiment with his friend Marry Currie. He cycling with Marry Currie and Says Catch me , and feel that when Marie Curie Comes around to them ,then time goes slow down . So In 1905 Elbert Einstein gave his concept of Special Theory of Relativity. In which he states that time were not absolute , it is Relative. Using his Own Concepts of Special theory of Relativity, Einstein Describe another Theory in 1915 which is popularly known as Einstein " General Theory Of Relativity " . in which he used his concept of " special theory of Relativity " (space & time) to Describe the " general theory of Relativity" . He Described that due to the Curved Path of space and time all the heavenly Body like planet Revolve around more heavenly body like sun. And this cause due to high value of gravitational force . And when this Gravitational force is very large as compared to inter-atomic repulsion of matter then all the particle or body revolved around each other will be collapse. So Albert Einstein Describe the Space Time curvature .And these all describe only in "General theory of Relativity " . But Einstein fail to explain what happened at singularity point, where all the matters will be existed in one point. And Using this General Theory of Relativity , many Physicist , and Scientist Describe Different Model . At that time one Problem is Rarely occur which is that Does Time have its Beginning or End ? How did the universe begin? Does Space have its edge? So these all question can be solved by using General theory of Relativity.

So here it is necessary to know about the universe , and necessary to know the beginning of the universe and. Only then we know about Black Hole.

So here we discuss different chapter from big bang to black hole. These chapters are.............

In first chapter we will Discuss about

1) Opinion about the Universe
   In Second Chapter we will discuss about

2) Enlarge of Universe or the expanding universe.
   In third chapter we will Discuss About

3) Black hole

4) In the fourth chapter we will discussed about current research regarding Black hole .

5) And in fifth chapter we will draw conclusion regarding the black hole.

So we will discuss these one by one so we wish to know about universe and Black Hole.

LITERATURE REVIEW

CHAPTER – FIRST

OPINION REGARDING THE UNIVERSE

Many opinions regarding the universe will be occurring at time to time in past. Idea about the universe firstly given by "Aristotle in 340 B.C , in his book On the Heavens ". He put forward the two opinion regarding the "Earth " was in a shape of round ball rather than the flat disk . First his opinion that Eclipses of the moon caused when earth comes between the sun and moon. .This was the first consequences of the Aristotle Regarding the Earth.

Second consequences of the Aristotle was that " Pole Star appeared at lower point of the sky when viewed from south pole i.e. pole star appeared for a few minutes in north region when seen from southern region of the universe . " Aristotle " , also estimate distance around the Earth which is around 300 hundred million stadia . it means that Aristotle indirectly measure the total distance of our Earth . it means that Aristotle said that a universe has its edge.

The Greeks (G.K) Philosopher also have third opinion that Earth must be of Round Shape . so Greek Philosopher and Polymath Aristotle also thought that Earth was always in a rest position (stationary) , i.e. Earth was placed at the center of the Universe and all other planets , stars and satellite Revolved around the earth . This theory is Known as " Geocentric theory ". So in this theory Earth is the center of the universe. And after some time " Claudius Ptolemy a famous , Astronomer ,and Natural Philosopher elaborate the idea of Aristotle in the 1st Centaury A.D (0 to 99 A.D.) ", gave a specific Cosmological model for Aristotle ideas . In his cosmological model Earth Was Placed at the center of the universe, and all the Planet and Stars moves around the Earth in an round path. And they took eight shell which were in the form of Planets and Stars ,Including Moon , Sun , Stars , and five Planet (Venus , Mars , Jupiter , Mercury , and Saturn) . So at that time there was no mention about other Planet, other galaxies, and other stars.

In Ptolemy model the outermost sphere is known as Fixed Star. And last sphere in this model is not clear to determine . Ptolemy model is useful after the Ptolemy to determine the position of Heavenly body . Ptolemy make an assumption that path Followed by moon that its path sometime twice the path of the Earth. This model is helpful to determine the increasing day and night in future days . But Ptolemy model Based on the assumption of Geocentric theory which predicted to be wrong in future . So this theory is not applicable for life time . And that time Ptolemy and Aristotle Model Totally Based on Accordance with Scripture. Because at that time all the people believes only on god but does not believe on real truth which is really occur . And those who used to stand against the geocentric theory got harsh punishment.

After some Time an simple model is given by " Nicolas Copernicus (famous Mathematician and

Astronomer) in 1514 . He Proved the Geocentric theory to be wrong. And give a idea that Sun was stationary, and all the object like Planet Revolved around the sun. This theory of Copernicus is known as Heliocentric Theory ". When Heliocentric theory was put forward by Copernicus then that time , same model was obeyed by " Italian Astronomer Galileo Galilei " . Galileo Galilei was the first person who made a telescope to see the Surface of Jupiter , He was also Studied the following physical phenomenon including , speed , velocity, gravity , freely falling body , relative inertia , projectile motion and also introduced the concepts of Pendulum . Using the pendulum concepts Galileo Galilei introduced the concept 'what is time '? After that Isaac Newton used the concepts of Inertia of Galileo Galilei and describes " 1st laws of motion ". Galileo Galilei also observes the phase of Venus, Saturn's rings and sunspots. So due to Copernicus theory and Galileo Galilei observation it will end of Aristotle and Ptolemy theory in 1609.

When Galileo looks to the Jupiter he observed that Jupiter is surrounded by several moon or satellite which is revolve around the Jupiter. Galileo was also to observed that moon of Jupiter is revolve around the earth in a complicated way. So he totally finished the theory which is given by Aristotle and Ptolemy.

And at the same time "German Astronomer " Johannes Kepler " start to give the review for Nicholas Copernicus theory and support to Nicholas Copernicus theory , he observed and says that orbit of different planet is not same which we seen from our eyes, so he directly says in opposite to Aristotle theory . so using own observation kepler gave the " laws of Planetary motion " which gives the direction of all the star and planets . Kepler also give the concepts of Elliptical motion of the earth. i.e. Kepler was the first person who said that the Earth have a Elliptical shape rather than round . Because in previous time this was said by Aristotle that Earth is an round shape rather than flat disk . so using this planetary motion concepts Kepler give the equation formula for Areal Velocity of Satellite. Therefore that time Kepler modified the Copernicus theory and introduced that all the planet moves around sun in an Elliptical path rather than in circular path.

And these all Explanation of Kepler was proved later by " Isaac Newton in 1687 " . When he publish his book " Principia " . This is the most important invention in classical physics. " Newton gave the three laws of motion in 1687 " . After in 1798 in according to " Principia Mathematica " Newton gave the laws of universal gravitational . Which describe the force of gravity . This is a force which helps to one body to keep with other body. He introduced the gravitational constant G , where the value of Gravitational Constant is equal to G = 6.674×10^{-11} $Kg^{-1}m^3s^{-2}$. And using the concepts of Gravitational force Newton also give the value of force of gravity on the surface of earth , which is equal to g = $9.8\ ms^{-2}$, so due to this force of gravity we continuously stay on the surface of earth. Before the Newton two people describe the force of gravity , firstly " Aryabhata" he explain why object will not thrown outside from the surface of earth , and Second one was " Brahmagupta " a sage who describe that force of gravity is an attractive force .

But they were not completely defining the force of gravity. That is why Newton name totally joined with gravitational force . And according to Newton this Gravitational force cause the moon to move around the earth , and cause to move all the planet to revolve around a sun in an elliptical orbit.

Newton gave also an assumption that Stars attract to each other with a force known as force of gravity ,so stars could not remain at rest , they continuously in motion, and observe that stars cannot fall at the same time.

After that a new thinker name Richard Bentley thought that there are a finite number of stars in a space. And also said that in an infinite space every point is regarded as the center of the universe. It is an rarely problem occur before the 20th century that nobody know that is universe Expanding or Contracting . Even those who realized that Newton theory of gravity expressed the universe is not remain at rest could not think that universe is expanding forever.

Another Assumption Regarding the universe is given by " German Philosopher Heinrich Olbers'' he suggested that in whole of the universe every point in a space is regarded as the center of the universe , which means that in the whole of the universe sky would be as bright as the sun have even at night. This consequence gives the thinking to researcher for sky and sun, and concluded the answer, if whole of the sky is bright as a sun does have then why we not see a star in a day? To avoid this consequences of Heinrich Olbers is that whole of the night sky should be bright as surface of sun only and only if the stars was not shinning forever. But we know that this is not possible. So answer of these all question can be taken out if we know that how our universe came into existence.

Review of Beginning of Universe

We know that the Beginning of the universe is Begin from Big Bang (theory which is already given). Many cosmologist gives the theory about the beginning of the universe , in past time all cosmologist states that the universe came into existence in a finite time . There are many arguments regarding the existence and beginning of the universe.

One of the argument regarding the universe is given by " Stephen Augustine " He pointed out in his book " The City of God " that all around us not exist forever and also civilization of our planet is not so old . And people activities also came in existenc when civilization begin so Augustine directly link the birth of the universe with the civilization of our planet.

So Augustine except that our universe made around 5000 B.C. This idea of Augustine was not followed by Aristotle and other Greek Philosophers because they were all thought that , the universe were goes on from past to present or future so on therefore no matter is that the beginning of the universe starts from the activity of human beings .That means Aristotle believed that there is no beginning or end of the universe exist. So when we study these all theory we could might to thing that , Does universe have beginning or end ? or Does universe goes on increasing or expand and contracting forever ?

The answer of these all questions is given by " Edwin Hubble in 1929 " . Edwin Hubble made a unexpected observation and states that " distant star i.e. star from other galaxies continuously move away from our solar system ,this means that our universe continuously expanding or goes increasing in size. So using these observation he states that after 10 to 20 million years ago all the matters are exactly at the same place i.e. singularity will exist. So when Edwin Hubble clear the Question of every person , which is that does universe exist same as forever ? He near the question which is that beginning of the universe ?

To answer this question Edwin Hubble suggested that there is a time before the Big Bang when everything of the universe placed at a same place i.e. it existed in a single point called singularity point . So in singularity point the universe was infinitesimally small and infinitely dense. And any event occur from previous time from big bang , that's there no means and time before the big bang is not affect the present time so that we will not considered. Therefore Edwin Hubble tells about Big Bang. So we may say that time had a beginning from big bang and earlier time from big bang has no means.

CHAPTER – 2

THE EXPANDING UNIVERSE

We know that we live in a planet earth which is a part of our Milky Way galaxy. our Milky way galaxy contains over two hundred billion stars , 8 planets , number of asteroids , comets ,satellites , dust particle and many more. So our solar system is known as milky way galaxy. For the previous time it was thinking by many people that our solar system is the whole universe. i.e. at that time many people think there is only one galaxy in the whole universe. But in " 1924 a American Astronomer Edwin Hubble demonstrated that there is number of galaxies exist in the whole universe. And after that Edwin Hubble determine the distance and Brightness of the other stars. He suggested that we can determine the distance of other galaxies , my measuring the distance or brightness of the stars. He suggested that If we measure the distance or brightness of other stars present at that galaxies , then we obviously measure the distance of other galaxies or distance between the galaxies . so we also find the distance or brightness of the other stars which is present in same or different galaxies and this measurement goes done when earth moves around the sun , but other galaxies are so far away from our galaxies that is why they really appear fixed therefore Hubble used indirect method to solve these kind of problem , so Hubble used this indirect method to measure the distance of galaxies . Therefore Hubble demonstrated that apparent brightness of the star depends upon two factor – first luminosity of star and second is distance of star from our galaxies. Hubble also suggested that there is certain types of stars which have same brightness , when they are very near to our solar system and we can measure their brightness .So this can be a consequences that there is stars in other galaxies which have also same brightness. So at that time " Edwin Hubble was work within distances to nine different galaxies ". We also know that our galaxy is the one of the galaxy from hundred thousand million galaxies , that can be seen by using telescope. Each galaxies itself containing some thousands million stars . And , our sun is a brightest star in the milky way galaxy of yellow color. One question arise from these all assumption which is that, Why stars appears to be very small ?

The answer of this question is that stars are so far away from our milky way galaxy, that is why they appear to be very small like a pinpoint , and we cannot determine there exactly size and shape , only we find their features , and color of light.

A discovery possessed by Newton ,which is that Isaac Newton pass the ray of light into a prism and determine that when a ray of light pass through a prism , the ray of light breaks up into number of light and gives the spectrum of light . We observe these spectrum also in Rainbow , So these concepts of Newton is also applicable in different reflection of stars . so we can observe different spectra for different galaxies and their stars and this is depends upon how the stars is hot and by the knowing glowing time of stars. And we know that each its elements of stars containing its chemical properties.

So by knowing the chemical properties of the stars be can find which elements presents in the missing stars and why they are missing from atmosphere.

At some time in 1920 when astronomers began to look at spectra of stars in others galaxies then that time astronomers find that a something different which is that others galaxies were moving away from him , and that time frequency of light waves from them was reduced , or shifted to red and this phenomenon of red shifted frequency is known as Doppler effect given by Johann Doppler. At that time Hubble spent a time to observe the spectra of galaxies and stars. At that time most of people were thought that other galaxies moving around in a randomly manner , and due to this the spectra of the wavelength firstly shift to blue and then shift to finally red , so that is why all galaxies appeared to be red.   in 1929 Edwin Hubble suggested that the color of the galaxies shifted to red were not randomly of permanently , but it is directly proportional to the galaxies distance from us . In simple word if galaxies were present at very large distance from us then it will moving away from us and vise versa. So this conclusion of Edwin Hubble suggested that our universe is not static or move randomly. So our universe expands in every second.

Newton, Hubble and many others should have suggested that this static universe would soon to start contracting after expanding.

Because suppose that if the static universe would to expand very slow then force of gravity would cause it eventually to stop expanding , and then to start contracting .

However this behavior of the universe could not be explain by using Newton universal law of Gravitational therefore behavior of the universe can be explain by using Einstein general theory of Relativity. So even when Einstein states about its general theory of relativity in 1915 he was not sure about universe was static or not ? so at that time Einstein modified his theory into a cosmological constant , in cosmological constant Einstein put a equation of antigravity , which is a type of force unlike to other force , and this antigravity force did not come from any particular source but comes due to the curve in space time i.e. built into the very fabric of space time. So Einstein cosmological model gave space time curvature and inbuilt tendency to expand. i.e. Einstein in 1915 gave the theory known as General Theory Of relativity which demonstrate that space and time are relative but not absolute . So Einstein modified the theory given by Isaac Newton , which states that Time and Space were absolute. And Einstein states that due the curve in space and time Fabric , heavenly body contain larger gravitational field around it , and smaller or less masses body contain small gravitational field around it, that is the reasons that all the planets moves around the sun. So Einstein cosmological constant make the balance to attraction of all the matter in the universe and make them so static universe.

Einstein Space Time Curve

After that German Physicist " Friedmann " explain equation ally to   general theory of relativity and support to Einstein general theory of relativity. Friedman gave the equation for general theory of relativity , and explain how the universe evolved in space time . friedmann suggested that whole of the universe looks identical in all direction , if we see the universe in whichever direction it looks like identical . So on the basis of this assumption of general theory of relativity Friedmann suggested that universe would not be static. In fact, in 1922, several years before   Friedmann model ,Edwin Hubble predicted exactly what friedmann found .The assumption that the universe looks the same in every direction is clearly not true in reality. For e.g.- , the others stars in our galaxy from a distinct band of light across the night sky called the milky way. So universe does seem roughly same in every direction. At that time the justification of our universe was not clear.

Now one of the landmark observation put forward by two American Physicist, " Arno Penzias and Robert Wilson in 1965 in Bell Laboratory of New Jersey " .They both of them make an microwave detector for communicating with orbiting satellite. At that time Arno Penzias and Robert Wilson when make a detector then both of them are very worried because their detector picking up more noise , And in starting they both of them think that any bird dropping on their detector, and this sound continuously detected, and this sound is very large when detector in a straight line.

So extra sound was continuously disturbing the atmosphere and detector were so noisy. So this sound was observed day and night, even when Earth was rotating around the sun. But after some time they both of them observed that this sound cannot be come from bird , but comes from outside the universe so this signify that radiation , continuously bombarded our universe. So this is most important invention at that time. And also at that Penzias and Wilson work on   Friedmann assumption .

Now at the   same time two another American Physicist , " Bob Dicke and   Jim Peebles work on this same assumption. They both of them were studied at Princeton University. They were both of them work on the assumption that how the early universe was very hot and dense, glowing white hot.   Peebles and Dicke argued that we see the glowing part of the universe because of that only just ray of light reaching to us. And however expansion of universe means the color of light directly shifted to red when we seen in microwave detector, and making this assumption finally they were observed that this assumption already done by Penzias and Wilson. So for this work they both them (Penzias and Wilson) get a noble prize in 1978. So all the assumption regarding the universe , that universe looks same in all direction , is verified truly. Now from these if all the universe looks like the same in all direction ,then one question arise from this ,which is that - If all the universe looks like the same then where is the center of the universe located ? this particular question arise at that time. Or this is the second assumption of the Friedmann's model. It is very interesting that universe looks same in every direction , but not in all points. In Friedmann model all the galaxies of the universe goes away from each other.

So in Friedmann model the distance between the galaxy is depends upon speed of the galaxies, this is most remarkable observation of the friedmann model .

This describe us how far from a different galaxies from one another. So this is similar to which is Hubble already found. So using the Hubble observation and success of Friedmann model , Friedmann is known as best for his work in the field of astronomy. So Friedmann found different types of assumptions and there solution , so these solutions are -:

1) In the first assumption Friedmann found that if universe is expanding sufficiently very slow so in this case gravitational force is very large , and this gravitational force cause to universe eventually stop to expansion. And in this case stars of galaxies moved toward each other and universe will be contract.

2) And second assumption of Friedmann model is that if universe is expanding so fast then in this case , universe will be expand forever.

3) And last or third kind of assumption is , if universe is expanding only just fast enough to avoid re-collapse ,   then in this case the separation of galaxies starts from zero or from singularity point. So in this case length of the galaxies   Expand forever .but speed at which galaxies move apart from each other is very slow although it never goes down to zero.

® Edwin Hubble (1889-1953), Alexander Friedmann (1888-1925)(L)

A remarkable observation of Fridemann is taken from first kind of assumption , which states that nor universe is infinite in space neither space have any boundary. He used general theory of relativity and states that if Gravity in any place is very large then at that place disorder will be occur, for e.g. if we start walk any point where gravity is very large then due to higher gravity we come back at the same point after some time . Friedmann used 4 dimensional analysis for represented any particle or object position , momentum and acceleration . Here the 3 dimensional is X,Y,Z and fourth dimensional is Time .so friedmann fourth dimensional is described time in a space. But 4th dimensional time is finite and present as a line with its two ends (beginning or end) . finite space can be given out only when we combine the General theory of Relativity with uncertainty principle which is later given by "Subrahmanyan Chandrasekhar " . We already discussed earlier that if force of gravity is very large in the universe and we start walk from any point , then after some time we goes back to starting point , from where we can start walking , but this is not possible , because for this you would need to travel faster than light only then we jumps back to previous position.

In other word Friedmann indicates to time travel . But here Friedmann gave the three assumption for our universe which we have discussed earlier , one question arise from these three assumption which is ? What which Friedmann model describes our universe ? will our universe eventually stop expanding ? Will our universe eventually start contracting ? Or will it expand forever ? To answer this question it is necessary to know current rate of expanding or contracting of the universe and its present average density. If the density is less than a certain grave value, determine by the rate of spreading out , then the gravitational attraction will be too weak to moratorium the expansion. If the density is greater then the grave value , gravity will stop the spread at some time in the future and cause the universe to re-collapse.

Now we can find out the present rate of expansion , by measuring the velocities at which other galaxies are moving away from us , using the phenomenon is known as Doppler effect . However we can measure distance of the different galaxies using indirect method, So we cannot find the distance of the galaxy accurately. So figure given by few astronomer for expanding of the universe is 5 TO 10 percent in every thousand million years. However this percentage of the spread out of universe is quite small. So if we add masses of all the stars that is contain in our galaxies or other galaxies then we can observe that this mass is even small for required to stop expansion of the universe. But we also know that our galaxies or other galaxies (whole universe) contain large amount of dark matter , which we cannot see directly , but which we know must be there cause because of the influence of its gravitational attraction on the orbits of stars and gas in the galaxies. Moreover most of galaxies are found in Clsusters , and we can guess the presence of yet more dark matter in between the galaxies in these clusters by its effect on the motion of the galaxies. When we add up all dark matter of the universe ,we still get only one tenth of the amount required to cessation the expansion. Yet we have another type of matter which we cannot detected , and which might still raise the average density of the universe up to serious value needed to cessation the expansion .So present proof ,gives the explanation that universe will continuously to expand forever. But don't bank on it . Even we all know that if the universe is going to re-collapse , it won't to do so far at least another ten thousands million years, since it has already been expanding for at least that long. So how the universe begun or is our universe can be expand ort?

This problem can be solved by knowing the concept of Big Bang.

BIG BANG THEORY

So Friedmann solve these kind of problem and suggested that there is a time in a past , before the 10 to 20 thousands million year ago , when distance between the neighboring galaxies is zero . i.e. all the mass of the galaxies exist in a point , called singularity point. At this point density of the universe , and curvature of space time is infinite. So singularity point is highly dense. Now at singularity point all the mass of the universe is concentrated into a single point so this point is absolutely dense , and due to this inter-atomic disturbance creates in a matter ,this is because of inter-atomic repulsion of matter , and as result explosion take place . This explosion in the universe is known as "Supernova" . So in this process whole of the matter disseminate in the universe. And this scattered matter combines with gas and form galaxies ,their stars , planet , satellites , comets , asteroid and many other particle in the universe. So this process when supernova occur is known as " Big Bang " . All our theory are made by taking the assumption that the curve in a space and time is smooth and flat.

So we can determine that the beginning of the universe begun from big bang. And existence of the time will be occur from big bang .

if there is event before the big bang then there is no use of that event , because predictability would break down at big bang. i.e. we could not find what happened before the big bang. Only we can find that what happened after the big bang . we can find the answer that , how our universe will came into existence , birth of the galaxies ,planets stars etc. but we could not find time before the big bang or what is exist before the big bang .

So we shall assume that the beginning of the universe occur from big bang , and before the big bang everything is zero. Many people at that time do not believe in big bang theory. After that In 1951 catholic church gave an model for big bang , we get this all from Bible. One of the most important theory regarding the big bang is known as Steady State theory.

This theory was given by "Hermann Bondi (Cosmologist) and Thomas Gold (Astrophysicist) , they both of them work on the development of radar during the world war. They both of them gave an idea that all the galaxies were moved away from each other and due to this a gap formed between the galaxies.

The steady state theory required the modification of General theory of relativity to allow the continues creation of matter. One of the most important predictions of this theory is that number of galaxies and other objects in the space is same wherever and whenever we look in the universe.

After sometime in late 1950 to 1960 , a source of radio waves in the outer space of the universe was taken out by Cambridge astronomer " Martin Ryle " . Other Cambridge Astronomer also suggested that this source of radio waves lies in outside of our galaxy, and also there is another source of electromagnetic waves in outer space of the universe. They source may be weak or strong. They observe that weak source lies more distance from us, and strong source lies very close distance from us . They also suggested that there is time in the past where all the sources lies in a same place and source may be very dense. Moreover discovery of microwave radiation by Penzias and Wilson in 1965 helps us to detect that there is a time exist in the past, when everything is more dense and infinite . So Cambridge Astronomer suggested that , steady state theory were wrong in prediction .

After some time in 1963, two Russian scientists, Evgeny Lifshitz (Soviet Born Physicist) and Isaak Markovich Khalatnikov (Soviet Born Physicist) suggested that there is a time when big bang occur, they were suggested that big bang may be idiosyncrasy of Friedmann's model on your own.

So Evgeny Lifshitz and Isaak Khalatnikov follow the Friedmaann model regarding the universe, and suggested that only one model will explain the singularity point at the time of big bang. So only one model exist which is known as Friedmann model. So Friedmann model gave the assumption of the real world.

So in Friedmann model all the galaxies moves away from each other. So in the past time all the galaxies are present at the same place. So in reality all the galaxies were not present at the same place but also very close to each other. So in present expanding universe there is not only occurring big bang but also as occur an contracting face of the universe. At the time of contracting universe, the universe will collapse and particles present in the galaxies not only collided but also they might have follow past and then away from each other, producing the present expansion of the universe.

So, At that time Lifshitz and Khalantnikov studied the model of the universe which is similar to given by Friedmann model of the universe. And this model gives us to irregularities and random velocities of the galaxies in the universe which is exist in real. They both of them showed that this model could be start with a big bang. And also showed that galaxies in our universe were not continuously move away from us. They both of them also realized that there is number of model which included in big bang but not included with singularity point ,only one model known as Friedman model which gives us singularity point in the universe.

Most important thing about the Lifshitz and Khalatnikov model is that, it is the most important model which describe the general theory of relativity correct , and showed that there is a time in the past , when everything of the universe is exist in a point is known as singularity point. It is clear from Lifshitz and Khalatnikov model that there Is a time in a past which is known as big - bang . So concepts of big bang is clear from this model . But now here one question is again left .Which is that if there exist a big bang then Einstein general theory of relativity predict that our universe should have big bang , a beginning of time ?

The answer of this question is given by "British Physicist , Roger Penrose in 1965 ". This answer is a completely different approach .

He used the way light cones, behave in general relativity , and introduced the concept that gravitational force is always attractive in nature . To show that a star that collapse under its own gravity is trapped in a region whose boundary eventually shrinks to zero size. This means that the matter of the star is shrink to zero size volume, so that the density of matter and the curvature of space time become infinite.

I.e. at singularity point, all the matter of star is in very dense position , Or according to Roger Penrose singularity contain within a region of space time is known as Black hole. Now all the persons at that time were satisfied from Penrose model . Penrose model states that any collapsing star must end in singularity, and universe is infinite in space. So we all clear that singularity of the universe will produced after expansion of the universe. Penrose also suggested that by crossing the singularity point we can travel time. But it is not easy , let we consider an example ,a person name X goes to his past ,and kill them his father name Y , now here time illusion will occur .because how will he be born when he finished his own father. So here time illusion will occur . so it is not easy to travel past or future in other word it is not easy to travel time . Einstein at that time suggested that it is possible to travel time through a wormhole or through Either. So all these concepts suggested that there was a time in a past when was everything of the universe place in a one point called as singularity point. So it is proved that beginning of the universe arise   from   big   bang.

Roger Penrose with Stephen Hawking

BIG BANG IMAGE

CHAPTER - 3

BLACK HOLE

A Black Hole is a region of a space time in which gravitational force is very strong so that nobody can be escape from it. Even light of electromagnetic radiation having velocity 3 million kilometer per second cannot escape from black hole region. We know that in our universe neither Can be travel faster than light ,so black holes don't even leave light.

The birth of black hole occurs due the death of stars. i.e. when a star is collapse due to its own gravity the whole mass of a star is concentrated into a single point called singularity point . so resultant , this point is quite unstable and that is why "according to Pauli exclusion principle the matter having the same atomic number cannot stay in an same atomic level. So due this strong inter- molecular repulsion is occur and this cause the explosion is known as "Supernova ". And because of this Supernova explosion all the particles of the universe diffuse from place to place and that is where the black holes originated. There are number of black holes in our universe. They are all in different shape and size. Also at the time of big bang ,many black hole were originated .They were all different shape and size .The black hole term firstly coined by " American scientist John Wheeler in 1969 " .At that time there are two theories about light. One was that light were composed of matter and second was it was made up of waves. We know that these both theories of light were correct, according to wave –particle duality of quantum mechanics. Because light can be regarded as particle as well as wave. Under the first theory that is ,the light was made up of waves, it was not clear from this theory that if light was made up of wave then how it would respond to a gravity (g). But according to second theory that if light were made up of particles, then one might except that it is affected by the gravity. One of the important assumption for this is given by an Astronomer and Natural Philosopher John Michell in 1783 .He was suggested that a star that was sufficiently massive and compact would have such a strong gravitational field around it so that light could not be escape from it. Any light emitted from the surface of the star would be dragged back by the stars gravitational attraction before it could get very far.

John Michell an Astronomer and Natural Philosopher (1724 – 1793)

Michell suggested that there are lots of star in the universe like this. Although we could not see all the stars because of the reason that ,stars are very far from us and light from that stars would not reach to us .And there are lots of objects in our universe which are very far from us and we cannot see them directly , however we would still feel their gravitational attraction. Such objects are what we now call black holes, because they contain black voids in space.

After some time similar suggestion was given by French scientist " Marquis de Laplace ". He gave the same idea which is already given by Michell.

So know how a black hole might be formed, this is the important question which we wish to know and answer of this question can be get by only if we know about the life cycle of a star. A star is formed when large amount of gas , mostly hydrogen ,starts to collapse in on itself due to its gravitational attraction. As it contracts, the atoms of the gas collide with each other, more and more repeatedly and at larger and larger speeds- then the gas heats up. Eventually the gas will be so hot that when the hydrogen atoms collide they no longer prance each other but instead converge with each other to form helium atom.

The large heat is produced in this overall process which is similar to the controlled hydrogen bomb, and this help the star to shine.

This additional heat not only increase the pressure of gas but also this heat is sufficient to controlled the gravitational attraction. Now at that time star will be stable for a long time because the heat produced from star produced the nuclear reaction which is balanced by gravitational attraction. Eventually, however after some time star will run out of its hydrogen and other nuclear fuels. This Is because if the star is more massive it is necessary to required more heat to balance its gravitational attraction. And if the star is very hot, it will use faster the fuel of itself.   Our sun has probably got enough fuel for another five thousands million years or so. One most of the more massive stars use up their fuel in little of time as hundred million years, this is much less the age of the universe. When star lose all of its fuel then they were start to cool and therefore they start contract. This is happened with stars, and this can be understand in the end of 1920s.

In 1928 an Indian student name '' Subrahmanyan Chandrasekhar work under the guide Sir Arthur Eddington who was expert on General Relativity. At that time Chandrasekhar work on the assumption that how big a star could be, and still separate itself against its own gravity after it had used up all its fuel. The idea of Chandrasekhar was put forward the answer that when the star become so small and matter particle become very near to each other. So Subrahmanyan Chandrasekhar play an most important role and solve the all problem regarding the star. Chandrasekhar use the Pauli Exclusion Principle of quantum mechanics for solve these kind of problem. So Pauli Exclusion Principle says that two matter particle cannot have the same position, velocity, and momentum in an atom. If particle have the same position, momentum and velocity then there will we born an uncertainty in position. So all the matter particle have the different position, velocity and momentum. This make them particle to move away from each other. Similar cases happened with star , matter present in an star move away from each other due to Pauli Exclusion principle and this make them star to expand. So after some time a star maintain itself with a constant radius.

Subrahmanyan Chandrasekhar an Indian – American Astrophysicist (1910 – 1995)

However , after some time Chandrasekhar realize that there is a limit of repulsion produced by the exclusion principle . Because the theory of relativity provides the limit to the maximum difference in the velocity of matter particle in the star to the speed of light . In the other word this means that when the star got sufficient dense, the repulsion produced from Exclusion principle would be less than to the gravitational attraction.   So Chandrasekhar put a limit for the mass of stars to be collapse. This limit of mass is known as Chandrasekhar limit.

So Chandrasekhar gives us to new idea for contracting and expanding of the universe. If a star of mass less than the Chandrasekhar limit then a star eventually to stop contracting and settle down to give us a White Dwarf with a radius very small comparable to few thousand miles and density of the white dwarf is approximately   equal to hundreds of tons per cubic inch. According to Pauli Exclusion Principle in white dwarf there is repulsion between the electrons in its matter. And in our universe we can find out a large number of white dwarfs. First white dwarf is known by us is Sirius B ,that we can find is orbiting around the star Sirius which is brightest star in the night sky. And distance of Sirius B from us is 8.6 light years.

A white dwarf Sirius B which is orbiting around Sirius A.

However a white dwarf is less dense as compared to neutron star and black hole .it's mass is equal to the mass of the sun and volume is equal to the volume of earth.

It is also realized by Chandrasekhar that there was another possible final state for a star with a limited value of mass of about one or two times to the mass of the sun, but much smaller than the white dwarf. I.e. when Supernova occur then another types of stars are formed They are known as Neutron star. In neutron stars there is repulsion occur in between the neutron and proton rather than the between the electrons. The neutron star having a radius very small comparable to 10 miles or so and density of neutron star is very high comparable to the hundreds of millions of tons per cubic inch. So neutron star is very dense as compared to white dwarf.

One problem occur at that this problem is what when a star is highly massive than the Chandrasekhar limit and lost all their fuels, So to solve this problem Chandrasekhar suggested that when a star highly dense then neutron star shrinks to zero size and then at that time laws of physics do not applicable to that region is called Black hole. So black hole is very dense then neutron star and its size are always shrinking to zero size. Before the Chandrasekhar   Nobody can be defined that star will collapse and shrink to zero size. Even Elbert Einstein himself written a paper in which he suggested that star would not shrink to zero size. So this can prove by only one person " Subrahmanyan Chandrasekhar ". For this commendable work, Subrahmanyan Chandrasekhar awarded by Noble Prize in 1983.

Chandrasekhar also proves that the exclusion principle could not stop the collapsing of a star more massive than Chandrasekhar limit. But problem is , to defined what happened to the star according to the General theory of Relativity , it was not solve until 1939 and solve after that by a young man " Robert Oppenheimer . He suggested that there is no any scientific consequence which cannot be observed by telescope. It is said that the Oppenheimer could take part in a project known as Manhattan project during the world war 2 and give a nuclear weapons power. After the year 1960, Astrophysics and cosmology was received a number of astronomical observations this will increase the discoveries in modern science.

And that time Oppenheimer observation rediscovered and extended by many scientists. The observation of the Oppenheimer work is that , the gravitational field of the stars changes the path of the ray of light and bend the space time curve. The light cones which indicates the path followed in space time by flashes of light emitted from their tips , are bent slightly inward near the surface of stars. This can be observed during the bending of light in eclipse of the sun, as the time of contracting of stars. Therefore the gravitational field at the surface of the stars is very high and the light cone gets bent inward more and more. This causally happened at event Horizon. So this makes light difficult to escape from strong gravitational field and as a result of this all the matter of the star is concentrated into a singularity. Therefore as result of this star in that point is visible like a dim particle when we see them through a telescope. According to the theory of relativity, nothing can be travel faster than light. Thus if the light cannot be escape from that gravitational field it means that neither can be escape from black hole gravitational field. So according to Oppenheimer this strong region is now called Black hole. And its boundary is known as Event Horizon, where the light cone bent inward more. If you want to know that how star is collapse inside a black hole , it is necessary to remember the theory of relativity which is states that time and space is not absolute. It is relative to the position of the body. Because each observer has own measurement of time from different space.   Elbert Einstein and Robert Oppenheimer worked together on the general theory of relativity. At that time Elbert Einstein discovered a equation E= mc^2 which is known as Mass Energy Equivalence. Oppenheimer gave the first nuclear weapons power in world war 2 and take part in Manhattan project. And use Einstein mass energy equivalence equation in this project. After that the outstanding work of Roger Penrose with Oppenheimer in 1960 to 1970 , changed all the misconceptions given by others . Roger Penrose suggested that a singularity point exist inside a black hole which is very dense and this singularity will be occur at the time of beginning of the universe and also occur when a star is collapse due to its own gravity. And at this singularity point all the law of science and physics cannot be valid.

Robert Oppenheimer (1904 – 1967)

In other word at singularity point law of science is fail to explain what happened at singularity point. On the other hand if a person or observer standing outside from a black hole, then black hole gravitational field did not affect on that observer or person. And a person / observer do not observe what happened at the region of black hole. Because neither light nor any other signal can reach them from the singularity point to that observer. This is the most important concepts of Roger Penrose and this is known as Cosmic Censorship Hypothesis. Cosmic Censorship Hypothesis states that if there exist a singularity in any place in the universe then those place is only a black hole. Where they are totally hidden from outside view by an event horizon .However we can see naked singularity according to general theory of Relativity. This is only possible when we avoid hitting the singularity and instead fall through a "Wormhole ", and come out in another region of the universe. This would be possibility to travel space and time, but this situation is highly unstable. Because this situation creates disturbance in space time. So we cannot see the singularity until be hit the singularity point. I.e. singularity point lies only in future and never in past. So here Roger Penrose tells about time travel. I.e. Roger Penrose tells that we can travel time through a wormhole without hitting the singularity point.

But it is not possible that we can travel a time without hitting the singularity point because , of the reason that when we goes near to the hitting point, the gravitational field of black hole attract to us in a very strong way. So Cosmic Censorship Hypothesis states that singularity always lies in the future, not in the past like the singularity of gravitational collapse, or entirely in the past like the big bang. Hence here is some possibility that we can travel time, i.e. we can travel in past .But this possibility to travel in past is very dangerous. Because let an example a person Name A, kill his father when go to past, so question arise that that person kill is father in past ,then how that person will take birth in future. So here time illusion is occurring, and this situation is quite unstable.

In 1967 an Physicist Werner Israel write a paper on the concepts of black hole and he write in his paper that a black hole, that is not rotating must be perfectly round or spherical .Its size depends upon its mass .And He was also write a particular solution for the general theory of relativity which has been found earlier in 1917 by "Karl Schwarzschild ". At first Israel found that black hole is only formed due the collapsing of bodies that were perfectly round and spherical. So we know that nobody can be perfectly round, so Israel said that the black hole should be behave like a ball of fluid .Israel's result had dealt only with the case of black holes formed from non-rotating bodies .On the analogy with a ball of fluid, one would except that a black hole made by the collapse of rotating body would not be perfectly round .so this theory of Israel have serious problem. After that in 1963," Roy Kerr had found a set of black hole solution of the equation of general relativity more general than Schwarzschild solution. The " Kerr " black holes rotated always with a constant rate , and their shape and size depends only upon their mass and rate of rotation .So Roy Kerr conclude that if a black hole rotation is zero , then the black hole was perfectly round in shape and this solution is identical to the "Schwarzschild solution " . But if the rotation was non zero i.e. not equal to zero, then the black hole bulged outward near its equator. So it was therefore natural to conjecture that a rotating body collapsing to form a black hole would end up in a state described by the Kerr solution.

After that in 1970 a research scholar name Brandon Carter , take the first step toward to apply conjecture.

He was showed that to provide an axis of symmetry to the rotating black hole the axis of symmetry like its size ,shape and spinning top ,all these depends upon only the mass and rate of rotation. Finally in 1973, "David Robinson " use the result of Brandon carter to show that the conjecture had been correct: such a black hole had indeed to be the Kerr solution. Therefore when gravitational collapse is occur, the black hole must be settle down into a state in which it could be rotating but not pulsating. Therefore black hole shape and size depends only on its mass and rate of rotation, and not on the nature of the body that had collapse to form it. In 1963, " Maarten Schmidt " an astronomer in California observatory found a faint star like object in the direction of source of radio -waves called 3C273 that is source number 273 in the third Cambridge catalog of radio sources, and he found that when gravitational field is strong around the black hole its color is shifted to red and that place the object is very massive and dense. This suggested that the red shifted was instead caused by the expansion of the universe. Which turn meant that the object is very far from us and that object is very bright and emitting a large amount of energy and we can also see them it from large distance.

Further an another explanation of the black hole is come in the year 1967, and this explanation was given by "Jocelyn Bell " . Bell states that some objects in the sky emitted regular pluses of radio waves .When Jocelyn bell and his supervisor " Anthony Hewish" stars works together than both of them thing that radio signal is due to because Alien civilization want to contact with us. And they can found four sources of radio waves including "LGM 1-4, LGM standing for 'little Green Men '.Here one question is arise which is that how we could find black hole ? The answer of this question is already given by " John Michell in 1783, that a black hole exerts a gravitational force on nearby star and also radioactive radiation is continuously emitted from black hole .When black hole overeats they will vomit those radioactive sources and finally we catch those radioactive radiations in radioactive source. And black hole is not really like a black body. However smaller the mass of a black hole larger they will glow and vise versa.

3.1   A SHORT LIST OF KNOWN BLACK HOLES

Stellar –Mass

NAME	CONSTELLATION	DISTANCE IN LIGHT YEAR	MASS IN SOLAR UNITS
Nova Mon 1975	Monocerous	2,700	11
Nova Persi 1992	Perseus	6,500	5
Nova Vul 1988	Vulpecula	6,500	8
Cygnus X -1	Cygnus	7,000	16
V404 Cygni	Cygnus	8,000	12
IL Lupi	Lupus	13,000	9
SS 433	Aquila	16,000	11
V4641 Sgr	Sagittarius	32,000	7
Nova Oph 1977	Ophiuchus	33,000	7

Galactic – Mass

NAME	CONSTELLATION	DISTANCE IN LIGHT YEAR	MASS IN SOLAR UNIT
Sagittarius A^*	Sagittarius	27,000	4,600,000
NGC-205	Andromeda	2,300,000	90,000
Messier-31	Andromeda	2,300,000	45,000,000
Messier-33	Triangulum	2,600,000	50,000
Messier-81	Ursa Major	13,000,000	68,000,000
NGC-1023	Canes Venatici	37,000,000	44,000,000
Messier-87	Virgo	52,000,000	3,000,000,000
NGC-3608	Leo	75,000,000	190,000,000
NGC-4261	Virgo	100,000,000	520,000,000

Here NGC-205, Messier-33 and Sagittarius star A is known as Intermediate Black Hole. And left of all is known as 'Super-massive Black Hole.

3.2   The Nearest Stellar Black Hole

A Black Hole is formed when highly massive stars explode due to their own gravity and this explosion is known as " Supernova ". But this is not happening near corner of our Milky Way galaxies so black hole is very far apart from us. The following are list of nearest stellar black holes-

NAME	CONSTELLATION	DISTANCE IN LY	MASS IN SU (Solar Unit)
1) Nova Monoce rotes 1975	Monocerous	2,700	11
2) Nova Persi 1992	Perseus	6,500	5
3) Nova Vulpecui l 1988	Vulpecula	6,500	8

4) Cygnus X-1	Cygnus	7,000	16
5) V404 Cygni	Cygnus	8,000	12
6) IL Lupi	Lupus	13,000	9
7) SS 433	Aquila	16,000	11

The Nearest Stellar Black Hole

3.3   The Nearest Super- massive Black Holes

There are number of Black Holes in the core of most of the galaxies ,they are very dense and energy of these Black Hole is entirely large .This types of black holes are millions of times massive than stellar black hole. Some of the galaxies have two or three black holes. Black hole never loses his mass. Some of the Super-massive Black Hole with its distance in light year from our galaxy is shown below:-

NAME OF BLACK HOLE	MASS OF BLACK HOLE IN SU	DISTANCE IN LY FROM OUR GALAXY
A) Sagittarius star A	4,600,000	27,000
B) NGC -205	90,000	2,300,000
C) Messier - 31	45,000,000	2,300,000
D) Messier – 33	50,000	2,600,000
E) messier – 81	68,000,000	13,000,000

F) NGC- 1023	44,000,000	37,000,000
G) Messier-87	3,000,000,000	52,000,000
H) NGC- 3608	190,000,000	75,000,000
I) NGC-4261	520,000,000	100,000,000

The Nearest Super-massive Black Hole

3.4   Moon as a Black Hole: -

Let us suppose that moon as a black hole, when we suppose moon as a black hole then we suppose that a group of hostile aliens passed through our solar system this will convert our moon as a Black Hole. So a body of mass million trillion tons and diameter of very small in mm can be converted into a moon as a black hole. Therefore we convert moon as a Black Hole if we convert, the mass of the moon million and trillion tons times to original masses of the moon and when its diameter is comparably small.

MOON AS A
BLACK HOLE

3.5   The Milky Way as a Black Hole:-

We know that in the center of a milky way Galaxy there exist an Black Hole is known as Sagittarius star A. It is very massive in size. That is why Sagittarius A^* is known as Super-massive black hole. The Sagittarius A^* has mass 2.6 times, more than to the mass of the sun. This black hole continuously attracts our milky way by a strong gravitational force. It is very brightest black hole in the universe and it emits continuously a source of radio wave in the Galactic center of the Milky Way. Sagittarius Star A is a super-massive compact object this can be proved by Two Physicist " Reinhard Genzel and Andrea Ghez " in 2020. That is why they both of them got a Nobel Prize in the field of black hole in the year 2020. Sagittarius A^* is 26,673 light year far from our planet Earth. We know that 1 Light year = 9460730472580.04 km

Therefore 26,673 light year = 26,673 × 9460730472580.04= 252346063895127500 Km

=$2.523460638951 \times 10^{17}$

Because we know that 1 light year = 9460730472580.04 Km

+ 26 light year = $2.459789922871 \times 10^{14}$ Km

+ 266 light year = $2.516554305706 \times 10^{15}$ Km

And the speed of Sagittarius A^* from which it continuously comes closer to Earth is 15,000 miles per second.

So from above calculation we can estimate that how is far a Sagittarius A^* from our planet Earth and also we will calculate when was our planet is totally collapse with this black hole.

Sagittarius A[*]

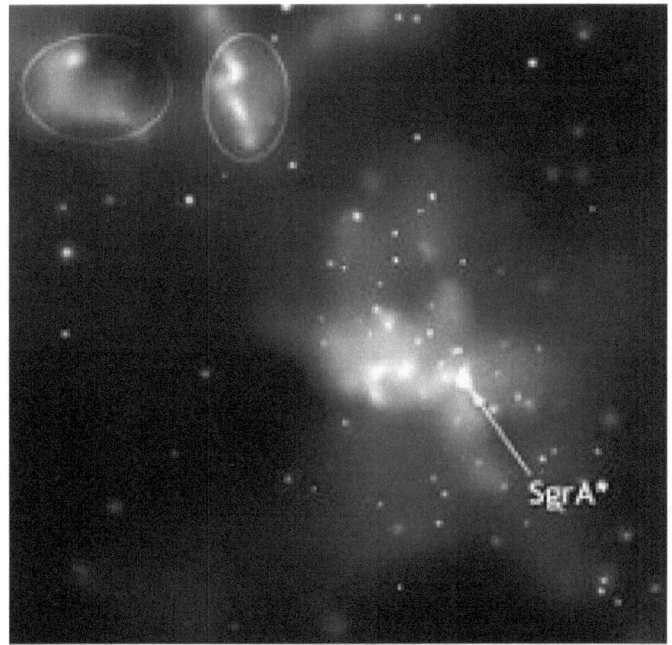

3.6   The Event Horizon and Singularity Point : -

We know that the Black Hole is a Region of space and time, where gravitational field is very strong that nothing can be escape from it. We know that nothing can be travelled faster than speed of light, but in Black Hole even light cannot be escape from it. There are a especially three parts of black holes: -

Singularity Point : -

Singularity point exist in the center of the black hole .And this point of the black holes comes after the Event Horizon. When all the matter of the black holes is concentrated into a single point , then that point of the black hole is known as Singularity Point . The Singularity Point is very dense due the reason that all the matter of the black hole is concentrated into a single point. And it is thought that by striking this singularity point we can travel time , because it is said that in singularity point there exist two curve point which join the different space and time .

Event Horizon : -

Event horizon exist in outside surface of the the black hole, and we can see the event horizon from outside through a telescope . I.e. Event Horizon is best known for its edge surface of Black Hole. Event horizon is a black spherical surface, with contain a very sharp edge in the space .Due to this sharp edge light becomes complled to bent at his surface.That is why it is generally said that in black hole even light cannot be escape from it.

Interior Space : -

Interior space is matted province in space time where space and time can get horrific dismember, Restrained ,expanded .And this place is very bad for time travel.

Event Horizon

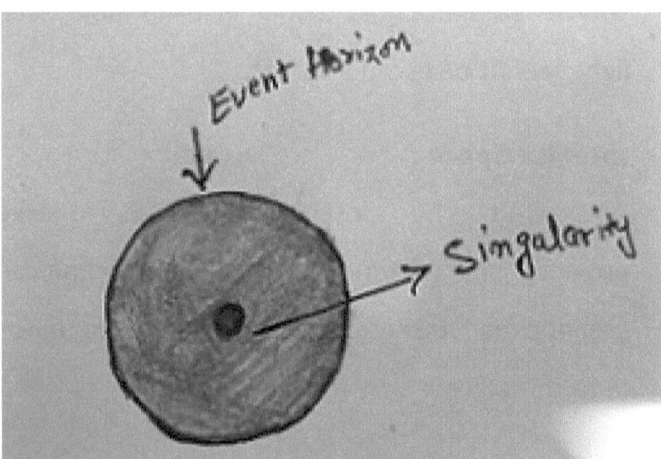

Singularity point

3.7   Gas Temperature of Black Hole

The temperature of a gas on the surface of a black hole is very high. It means that it get very hot and emit continuously light. This light is visible when we see a black hole through a telescope. It is felt at that time when Black Hole eats everything around it ,and eats it so enough that it feels overeat and start vomiting. At that time black hole release energy and this energy can be seen in the form of light. Here inside the black hole gas is much heated due the reason that the atoms are continuously collide with each other. And the matter which is far from black hole do not heat up so, this matter is much cool as compared to matter present inside the black hole. The atom which is very closer to the black hole moving with speed of millions of kilometer per hour, so gas of that matter is thousands degree times much more than the outer matter which is cooled.

Black Hole as emitting Ray of radiation or light

3.8　　　Time Dilation near the Earth and Black Hole: -

The general theory of Relativity which is described by Elbert Einstein in 1915 gives the some unusual predictions, which is verified by experimenantly by Elbert Einstein and states that two people experience a different times, where one people is standing on the surface of the earth and second person will be go on space .Here time is different for both of them　because the gravitational field on both the place is different . An another example for this type of problem is – let a person A go to space with the speed of light and another person　B is stay on the surface of earth. So when a person A comes back to earth he sees that person B looks very aged. This all happened because that when a person A goes to space with the speed of light, then for that person time goes slow down. Because nature do not break their own laws. So according to Einstein, time is Relative but not absolute, and depends upon the position of the body. Einstein also concludes that space and time also depends upon their gravitational attraction. So when two observer, absorb different time in different space, this is called Time Dilation. So time dilation due to the earth Gravitational field is given from the formula: -

$$T = t\sqrt{1 - \frac{2GM}{RC^2}}$$

Where T = Time measure from Planet Earth

t = Time measure from space

M = Mass of the planet in kg

R = The distance of the observer who is far away from the Planet Earth.

So, using this formula we can calculate the difference of the ages of two people A and B i.e. we can find that what is the age of the person who goes to space and also the age of the person who can stay on the surface of earth.

And also time dilation, near the Black Hole is experienced due to the strongest gravitational field around the black hole, however this gravitational field is very large than earth gravitational field, so according to general theory of relativity, presence of space and time inside the black hole did not determine. This means that no space and time exist inside the Black Hole, However space and time exist for black hole at Event Horizon i.e. at outer edges of the Black Hole. And we can find the time dilation at Event Horizon using the same formula as we discussed above.

3.9   Accretion Disk Temperature Of a Black Hole : -

When a body moves toward a black hole then that body will move with larger speed due to high gravitational field around it. So kinetic energy of the body gradually increases and as a result of this the Accretion disk of the black hole moves with larger speed and thus large amount of heat is released from accretion disk of the black hole. So we can say that temperature of the Accretion disk is very large. And it is naturally to observed that accretion disk of every black hole continuously moves a larger speed. And this speed of the accretion disk depends upon the factors that how massive a star is .And the temperature of a accretion disk can be calculated by using the formula : -

$$T = \frac{2GMm}{3kR}$$

Here,

G = value of Gravitational field

M = mass of Black hole

M = mass of the body or particle

K = Boltzmann constant

R = distance between the black hole and other particle.

Black Hole Accretion Disk with Radiate Heat

4.0 Distortion near the Black Hole: -

In the place of black hole space time will be distorted, that means that space and time gives the different value then as compared to other place. This is the importation thing happened to that place ,which is that time will be distorted . According to general theory of relativity we cannot travel backward or past from present time .The chances for travel in past is very rare and dangerous. It is generally defined in special theory of relativity in 1905 that the for different person the position of the particles will be different and depends upon its position ,momentum and velocity, although in previous time frame of references is defined by Newton , but he was said that time will be absolute and not changed when we measure the time for a body from different position .Therefore Newton explain that phenomenon which is generally occur in our planet and fail to explain why the time will be change when we go to other planet or fail to explain why inside a black hole there is no exist a time. It is defined by Elbert Einstein in his theory Special theory of Relativity in 1905 .So for different positioned person the space and time will be different. For example when a body moves with speed of light then for that body the time will go slow down as compared to other body. Because of the reason that nature do not breaks their own laws. We can see the similar case when we go near the black hole ,when we go near the small sized Black hole then in that case the time will be go slow down but when we go near the super-massive black hole then there is no existence of time inside the super-massive black hole . Because of the reason that inside a super-massive black hole there is large amount of gravitational field around it, and that is why the speed of the matter will be goes slow down which is near the black hole. Because according to general theory of relativity gravity will be effect on the mass of the body .Heavy will be the body large will be its gravitational field and lighter will be a body smaller will be its gravitational field. So inside a black hole there is no existence of time and this create the distortion near the black hole.

Distortion near the Black Hole

4.1 Gravitational Field near the Black Hole: -

The gravitational field near the black he very strong that nobody can be escape from it. We know that how space time position of the body depends upon its frame of reference. So we now that the gravitational field near the black hole is very strong so if we wish to see what happened at event horizon then it is necessary that we put our instruments like satellites at a larger distance from event horizon otherwise strong gravitational field of black hole attract satellite in a strong way . Because we know that even a light can't be escape from a black hole.

So we only see what happening in a black hole at a large distance. Therefore at a large distance we can calculate the speed, time, brightness and other factors of the black hole. So it is concluded that inside a super-massive black hole there is no existence of time.

This curve shows the effect of gravity on space time.

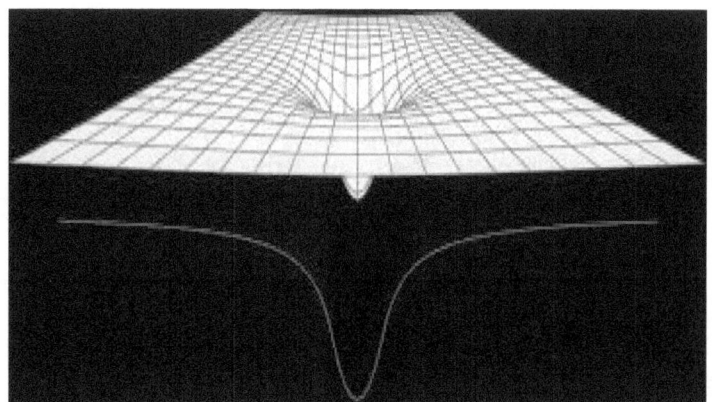

Black hole with its high Gravitational field

4.2   Tidal Force on the surface of Black Hole: -

A tidal force is defined as the difference in the strength of gravity between the two points. Tidal forces is naturally to be observed in the gravitational field of the moon , i.e. the gravitational field of the moon produces a tidal force on the surface of the earth cause the earth defeaters . This tidal force is also responsible for tides in the surface of the earth with several meters. And is responsible for larger tides in the liquid oceans. If the tidal force is greater than the body cohesiveness then the body will be disrupted .Same situation is occur when a person is fall into a black hole .When a person is fall into a black hole they experience a tidal force. The differences in the acceleration of the black hole due to tidal force are thousands time the earth gravity. The person will be pulled with large distance due the effect of tidal force.

Tidal Force on the surface of Black Hole

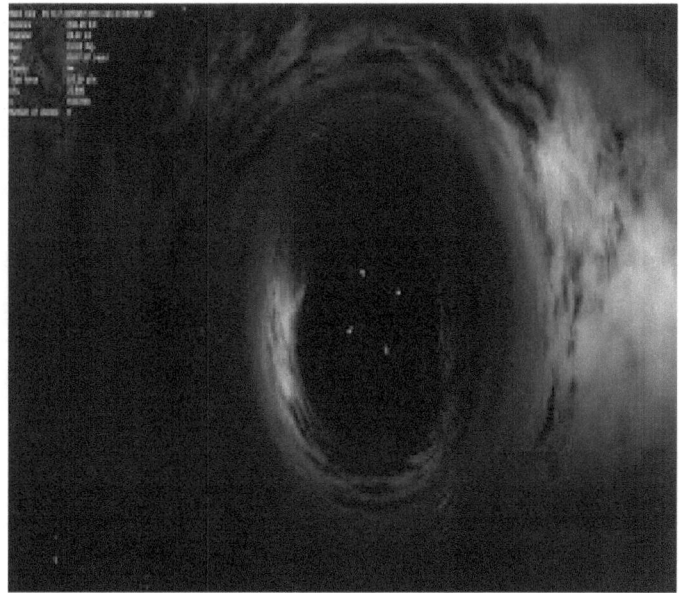

4.3 A star Collapsing into a Black Hole : -

We know that the birth of Black hole is cause due the death of star. So when a star is collapse due to its own gravity it contracts it into a singularity point and form a black hole. And at a time of collapsing a large amount of explosion is occur and this explosion of stars is known as " Supernova ". So we can say that when a star is collapsing in a supernova process after that all the matter of this star is collected into a single point and form a black hole. And other material is dissected into the universe to form another stars, comets and asteroid. The supernova process of the stars is occur due to the reason that at the time of supernova the surface of the star is so dense and due to this increases in density the surface of the star become so hotter and hotter and as a result of this supernova process of the star occur.

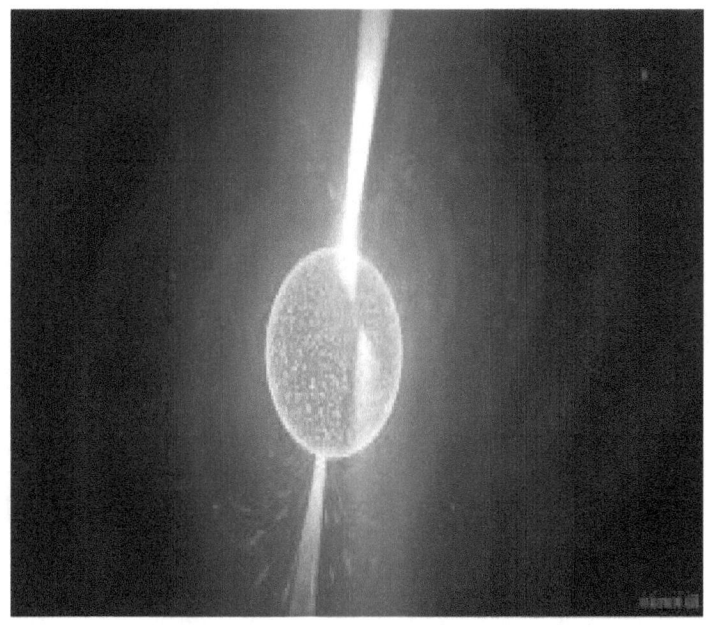

Star Collapsing into a Black Hole

4.4   Falling into a Black Hole and travel time: -

It is really not possible that a person goes easily inside a black hole and travel time. let us we suppose that a person will try to go inside the black hole .So when a person go near to a black hole they observe a large amount of gravitational field around it and this increase the kinetic energy of that person, as a result of this person move with high speed and this increase the heat around that person. So after some time that person will be goes inside a black hole and travel time. According to general theory of relativity we can travel a time through a Wormhole. But it is not easy to travel a time in a past or future. Because according to general theory of relativity if we wish to travel time through wormhole then we have to go through a lot of gravitational field. And this even causes death of that person. So it is believe that black hole singularity point is the only one place from where one universe is joint with other. In other word, through black hole singularity point we can travel time. But this is not easy to travel time through black hole. Because we already discussed that in black hole singularity point there is a lot of gravitational field which produce amount of heat in the core of black hole. That's why we go there to challenge death accordingly. And also when we travel time then it create time illusion .for example , let a person go in past and kill his father, so here one question is arise that if that person kill his father when go to past then how that person will take birth in future. Therefore here time illusion will be happened.

4.5   What Happened inside a Black Hole :

When you go near the Event Horizon of the Black holes then at event horizon time illusion and many other optical illusions will be occur. And it is also concluding that Black Holes generate continuously a large amount of tidal and gravitational force around it. And when you go very close to Black Hole then gravitational force and tidal force of Black Hole attracted you by at a large extent. And will even keep your duds ragged. So general theory of relativity predict that we can travel a time when we go through a black hole. But we already discussed that this is not easy to travel a time through a black hole. So inside a black hole there is a large amount of gravitational force, tidal force and many other electromagnetic radiations. So when we go inside a black hole we have to go through all of them. And when you reach at event horizon then nobody can escape you from gravitational force. Because in event horizon gravitational force is very large. So finally you permanently hanged in event horizon and after some time this gravitational force of black hole attract you to inside a black hole. You only escape from that gravitational force if your speed is faster than speed of light. But we know that this is not possible.

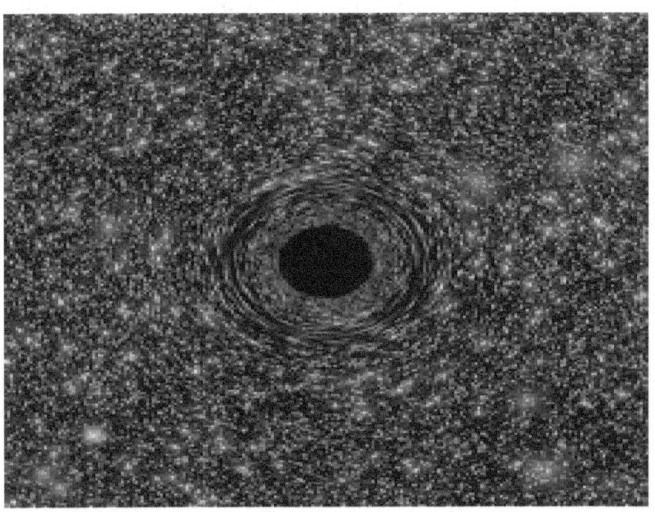

Images when we looks inside a black hole using telescope

4.6 Working into a space time near the Black Hole: -

According to " Einstein General theory of Relativity " at the near of Black Hole the space will be greatly distorted, some of the people call this as 'warping 'of space. When the Einstein theory of relativity is studied by "Schwarzschild " in 1916, then he found the mathematical formula for describing the warping of space in the Black hole. So space time curve of the black hole generally bent in shape. And this is the reason that the light cone will be bent, when goes to the surface of the black hole. So due this curve of space and time in Black Hole, Black Hole even never left to light.

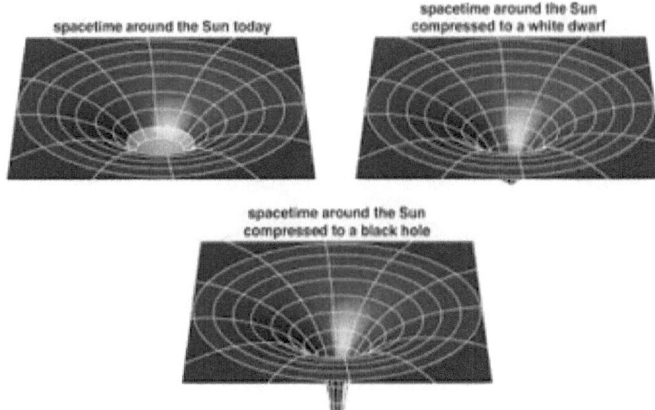

Space Time curve of a Black Hole

CHAPTER – 4

Current Research Regarding To Black Hole

In the year of 2020 three Physicist named, " Roger Penrose, Reinhard Genzel and Andrea Ghez done an remarkable work in the field of Black Hole. Roger Penrose starts is work in the year 1965 in the field of Astronomy and Cosmology. Than at that time Penrose propose the cosmic censorship hypothesis .This hypothesis states that the black hole contains its singularity point in its center where all the mass of the matter is concentrated into a single point. So Roger Penrose suggested that there exist a singularity within a black hole and at the singularity point the black hole is very dense. Or this density of the singularity is like a big -bang. It was also suggested by Penrose that at singularity point all the laws of science will be fail to explain that what happened at that point. Any person who is stay outside from singularity will not be predicted that what happened at singularity. But observer stay outside from Event Horizon will be predicted that what happened at Event Horizon.

But don't know completely what happened at Event Horizon. Because of the reason that at that place neither light nor any other signal will be reach from singularity. So this is the Remarkable opinion of Roger –Penrose. Roger Penrose also suggested that the singularity of Black hole will be produce due to gravitational collapse of the stars and this singularity will be occurring in the place of black hole and at the time of Big Bang only. And outside view of the black hole is totally hidden by its outer edge known as Event Horizon. So Roger Penrose used his hypothesis which is known as cosmic censorship and said that observer who is stay outside will not predict that what happened at singularity point. If a person wants to know what happened at singularity point then it is necessary to hit the singularity point. Roger Penrose also suggested that there is some solutions of general relativity in which is possible for our astronaut to see a naked singularity. The astronaut may be able to avoid hitting the singularity and instead fall through a " wormhole" and comes out in another region of the universe. This means that Roger

Penrose tells about to travelling space and time. I.e. Roger Penrose tells about time travel. But Roger Penrose also suggested that it is not easy to travel a time because at that time situation will be unstable and creates a time illusion.

So he conclude that without hitting the singularity point an astronaut cannot travel through space time. So this means that Penrose suggested that a singularity will be lying in Future but not in past. So " cosmic Censorship Hypothesis " states that an singularity will be lie entirely in future like the singularity due to a gravitational collapse or lie entirely in past means at the time of beginning of universe i.e. Big Bang singularity. So Roger Penrose states that when we want to travel in past then it creates a time illusion for example ,if a person kill his father when go to past , here one question is creates which is that if that person kill his father when go to past then how will they get birth in future. A black can have in any shape, and its size not even be fixed. So Black Hole size pulsating continuously. So from this Roger Penrose Predicted that General theory of relativity gives the complete information regarding the Black Hole. Roger Penrose use mathematical methods to prove that Black holes can directly work on the consequences of general theory of relativity. Einstein himself did not believe that Black Hole exists. In January 1965 Roger Penrose starts work on the consequences of black hole and prove that black hole really existed and Penrose describes really it very well. So this work of the "Roger Penrose is the most important contribution in the field of " General Theory of Relativity ".

Therefore for this outstanding work, The Royal Swedish Academy Of Science has decided to give a Noble Prize in the field of physics to Roger Penrose in the year 2020. So half part of this Nobel Prize will be given to Roger Penrose

And next half part of this Nobel Prize will be given to Two Physicist named Reinhard Genzel and Andrea Ghez. Reinhard Genzel and Andrea Ghez both of them done an outstanding work for a Milky Way Galaxy, Black Hole named Sagittarius A^*. Both of them were working together from 1990s. Both of them proved that the super-massive Black Hole name Sagittarius A^* present in the core of the Milky Way galaxy. By using world's largest telescope, Reinhard Genzel and Andrea Ghez developed a method to see through the huge interstellar gas and dust to the center of the Milky Way. I.e. they were used mathematical method to prove that Sagittarius star A (which is a super-massive black hole) lays inside the Milky Way galaxy.

When Roger Penrose suggested that the black hole is formed due the prediction of General Theory of Relativity then Reinhard Genzel and Andrea Ghez will be discover the super-massive compact object at the center of the galaxy. Therefore Penrose have get the Nobel prize due the Reason that Penrose suggested that   A Black Hole will obey the Prediction Of General Theory Of Relativity and

Genzel and Ghez get Nobel Prize due the reason that they both of them discovered the super-massive compact black hole at the center of Milky Way Galaxy. So both of them Reinhard Genzel and Andrea Ghez is awarded by the half of the Nobel Prize in physics in the year 2020 , and left half part of the Nobel Prize will be given to Roger Penrose .So Andrea Ghez become the 4th women in the world who take a Nobel Prize in Physics.

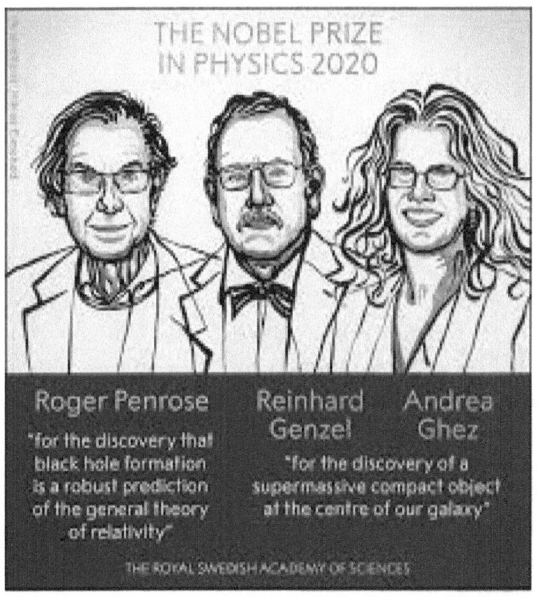

**Nobel Prize winner in the field of Physics
in the year 2020**

Roger Penrose Born -18 Aug 1931 (Mathematical
Physicist, Philosopher of Science)

Andrea Ghez Born 16- June 1965 (Astronomer and Professor)

Reinhard Genzel Born -24 march 1952 (Astrophysicist)

CHAPTER – 5

Conclusion and Future Application :-

Black hole is a space in a region where gravitational field is very large, that nothing can be escape from it .Even light having a speed three million km per second cannot be escape from black hole. A Black Hole is formed when a star is collapse due to its own gravity. In other words the Birth of Black hole is cause due the death of a star. Everything of Black hole can be defined by using " General Theory of Relativity " .And is verified by Roger Penrose in 2020 that Black hole can be work according to General theory of Relativity. Black Hole exerts strong Gravitational Force, Tidal Force and other Electromagnetic radiation on other bodies with a larger speed. There are many types of Black Hole in whole universe . some of the black hole are small in size and others are large in size and that larger sized black hole is known as super-massive Black hole .And that super-massive Black Hole are very dense this means that heat and temperature of that black hole will be very large .Generally we divide Black Hole in two Categories as: - 1) stellar mass Black Hole and 2) Galactic mass Black Hole. And these two categories of black hole contain as a some of intermediate −mass black hole and others as a super-massive Black Holes. One of the most important future application of Black Hole is that we can determine 5th Dimension when we go inside a Black hole through singularity point. We know that according to general theory of relativity their exist only a four dimensional namely one dimensions ,two dimension and three dimensional and fourth dimensions is time , which exist in generally. But it is predicted that when we go through a black hole we can also travel in a 5th dimensions ,which is the dimensions for travel through space and time .But this exist only in prediction .Because it is not easy to go through Black hole and travel through space and time.

References: -

1. A Brief *Ferguson, Kitty (1991). Black Holes in Space-Time. Watts Franklin. ISBN 978-0-531-12524 - 3*

2. *Hawking Stephen (1988 A brief History Of Time Bantam Books, Inc. ISBN – 978-0-553-38016-3*

3. A Brief History of Time J.B. Hartte; S. W. Hawking; ''wave function of the universe ''. physical Review D.28 (12) : 2960

4. Mckie Robin . '' A brief History of Stephen Hawking .cosmos .Retrieved 13 June 2020

5. Gribbin ,John ;white ,Michael (1992) .Stephen Hawking : a life in science ISBN 9780670840137

6. History of time An interacting adventure

7. The History of Everything : Stephen .W. Hawking ISBN 81-7992-591-8

8. ''Stephen Hawking (28 Feb 2016) The theory of everything: The origin and fate of the universe. Phoenix Books , special Anniv, ISBN 978- 1- 59777-508-3

9. ''Hodge; Johe .c. (2012) .The theory of everything: Scalar potential model of the Big

Bang and the small .p.p.1- 13, 99. ISBN 9781469987361

10. "Newton sir Isaac (1729). The Mathematical Principle of Natural Philosophy " 11.p.255.

11. "Abraham pais (23 Sep 1982) : The Science and life of the Elbert Einstein : oxford university Press ISBN 978-0-19-152402-8

12. Horowitz, Garlf. " The origin of Black Hole Entropy in String Theory ":ag –qc / 9604051

13. Greene , Brian R ; Momican , David R; Strominger , Andrew (1955) , " Black Hole

14. Condensation and the unification of string Vacua " Nuclear physics B : 451 (1- 2) : 109-120

15. Bland – Hawthorm ; J :2013 Black Holes / 4979088

16. Crothens , S; Super-massive Black Hole at Sagittarius A^* ,2019

17. Gillesesen , S ;General .R ; Elsenhauer , F; new observation detail of milky way galaxy ,Big Black Holes , December 9,2008

18. Gillessen , S , supermassereiches Schwarzes Loch Vorwandelt Gaswalke in " spaghetti , 20 July 20013

19. Hawking , S.W ; The theory of everything ,the origin and fate of the universe (New Millennium Press , Beverly Hills ,C.A 2002)

20. Hawking ,S.W. information Preservation and weather forecasting for Black Holes , 22 Jan 2014

21. Black Hole Math: space math @ NASA; Feb 2019,Sten,F.Odenwald@nasa.gov. ;http://spacemath. gsfc.nasa.gov

22. Death Spiral around a black hole – Hubble discovery ; http://hubblesite.org/newcenter/archieve/released/2001/2003

23. *Chandra Observatory Detects Event Horizon - http:Chandra.harvard.edu/photo/2001/blackholes/*

24. *Ask the Astronomer:87 FAQs About Black Holes : http:www.astronommycafe.net/qadir/abholes.html*

25. *Imagine the universe : Black Hole FAQs ; http://imagine.gsfc.nasa.gov/docs/ask_astro/black_holes.html*

26. *New evidence for Black Holes from NASA;*
    http://science.nasa.gov/headlines/y2001/ast1
    2jan_1.html

27. *A   trip   into   a   black   hole   ;*
    http://antwrp.gsfc.nasa.gov/htmltest/rjn_bht.ht
    ml

28. *Beyond Einstein :from big bang to black hole*
    ;OCT2012   ;   http://universe.gsfc.nasa.gov/
    ;FS -2002-12-050.GSFC

29. *Wald 1984 ,p.p – 299-300*

30. *Wald RM, (1997); ''Gravitational Collapse and*
    Cosmic censorship ''. In Iyer , B.R ,Bhavoal ,
    B(eds) Black holes, Gravitational Radiation
    and the universe, Springer ,p.p.69-86 ISBN
    978-9401709347

31. *Hamilton, A,''Journey into a Schwarzschild*
    Black Hole '', 28 June 2020.

32. *Schutz , Bernard F.(2003), Gravity from the*
    grounded   up   ;Cambridge   university
    Press.p.110 (2 Dec 2016)   ISBN 978-0-
    521-45506-0

33. *Daries, P.C.W. (1978). ''Thermodynamics of*
    black holes, report on progress in physics 41
    (8):1313-1355.

34. Hawking .Stephen, Penrose Roger (1996). The Nature of space and time, Princeton University Press. ISBN 978-0-691-03791-2

35. Melia ,Fulvio (2003) The Black Hole at the center of our galaxy Princeton University Press ISBN 978-0-691-09505-9

36. Melia ,Fulvio (2003) the edge of infinity ,super-massive black holes in the universe ,Cambridge University Press ,ISBN 978-0-521-81405-8

37. Throne ,kips (1994) Black Holes and Time Warps .Norton .W.W. and company ISBN 978-0-393-31276-8

38. Pickover Clifford (1988) .Black Hole :A Travelers Guide ,John and sons ; ISBN -978-0-471-19704-1

39. Carroll, Sean M.(2004), space time and geometry .Addison Wesley ISBN 978-0-8053-8732-2

40. Carter, B (1973) ''Black Hole equilibrium states '' in Dewitt , B.S; Dewitt .C.(EDS) .Black Holes.

41. Susskind , Leonard (2008) ,The black hole War : my battle with Stephen Hawking to

make the world safe for quantum mechanics ;
ISBN 978-031601670

42. Chandrasekhar , Subrahmanyan (1999)
mathematical theory of Black Holes ;Oxford
University Press ;ISBN 978-0-19-850370-5

43. Wheeler ,J .Craig (2007) ;Cosmic
Catastrophes (2nd Edition) Cambridge
University Press ;ISBN 978-0-521-85714-7

44. Gallo, Elena, Marolf, Donald (2009),
"Resources Letter BH-2 Block Hales".
American General of Physics.

45. Review Paper on Hughes Scott A (2003),
"Trust but verify the case of Astrophysical
Black Holes ".

46. Black Hole Physics : Fundamentals theories
of physics ,96.1998 ;ISBN 978-0-7923-5146-7

47. Hawking .S.W ;Ellis ,G.F.R. (1973) Large
scale structure of space time ;Cambridge
University Press ;ISBN 978-0-521-09906-6

48. Prince , Richard ; Creighton ,Teviet (2008)
"Black Holes " 3 (1) : 4277

49. Wald, Robert M. (1984). General Relativity (
University of Chicago Press) ;ISBN 978-0-
226-87033-5

50. Taylor Edwin F, Wheeler John Archibald (2000) ; exploring Black Holes .Addison Wesley Lengman ; ISBN 978-0-201-38423-9

51. Frolov, Valeri P; Zelnikov Anderi (2011); introduction to black hole physics; oxford university press. ISBN 978-0-19-969229-3

52. Wald, Robert M (1992); Space Time and Gravity: the theory of the big bang and black holes.

53. Misner Charles ;Thorne ,Kip S; Wheeler John (1973) , Gravitation .W.H. ISBN 978-0-7167-0344-0

54. Melia, Fulvio (2007) .The Galactic Super-massive Black Hole. Princeton University Press ;ISBN 978-0-691-13129-0

55. Clery D (2020) ''Black Holes caught in the act of swallowing stars ''science'' 367 (6477):495

56. Abbott ;B.P ;et al (2016) observation of Gravitational waves from a Binary Black Hole Merger 116 (6) : 061102

57. Siegel ,Ethan '' Five Surprising Truth about Black Hole from LIGO '' 12 April 2009

58. Event Horizon telescopes, (2019) "First M87 Event Horizon telescope results. The shadow of the super-massive Black Holes : The Astrophysics Journal 875 (1)

59. Landau, Elizabeth (10 April 2019) "Black Hole images makes History " (NASA).

60. "Ripped apart by a Black Hole ESO Press " Release on 21 July 2013 ; Retrieved 19 July 2013

61. Droste ,J (1917) " on the field of a single centre in a Einstein's Theory of Gravitational ,and the motion of a particle in that field ; Royal Academy Amsterdam 19 (1) 8 197-215 ;Archived 16 Sep 2012

62. Eddington, Arthur (1926) .The internal constitution of the stars. Sciences 52. Cambridge University Press .p.p.233-40-ISBN 978-0-393-31276-8 ; 11AUG 2016

63. Venkataraman, G. (1992). Chandrasekhar and his limit ;University Press.p.89; ISBN 978-81-7371-035-3 ; 11 AUG 2016

64. Hawking ,S.W. (1974) "Black Hole Explosions in nature (248 / 5443): 30 -31

APPENDIX / APPENDICES

1. Accretion Disk: - According to general Relativity an Accretion disk is a outer disk of the black hole which continuously move with a larger speed. This Accretion Disk Contain large amount of Dark Matter including dust particles and gaseous form.

2. Big Bang: - The theory of Cosmology in which the expansion of the universe is happened due the large explosion.

3. Chandrasekhar limit: - Chandrasekhar limit is the limit of a star in which different star make a White dwarf, Neutron star, and Black Holes during the process of supernova explosion. This limit is equal to the 1.4 times to mass of the sun, Therefore Chandrasekhar Limit = 1.4 times to the mass of the sun.

4. Cosmological Constant : - This is equation added by Elbert Einstein in General Relativity to account for an apparently non expanding universe , and this can be further rejected by Edwin Hubble to performed that this have no needs .

5. Cosmology: - This is the Astrophysical study of the History, structure, and dynamics of the universe.

6. Dark Energy: - The Residual energy in empty space which causing the expansion of the universe to accelerate. Einstein cosmological constant was a special form of dark energy.

7. Dark Matter: - This is the matter which is present in the core and outer edge of black holes. This matter have large amount of kinetic energy and also contain large amount of heat.

8. Doppler Effect: - An observer receives sound and light from bodies moving away from with lower frequency and longer wavelength than emitted with red shift and body moving toward with her with higher frequency and shorter wavelength. This shifts in frequency increases as the speed of the body increases. This phenomenon is known as Doppler Effect.

9. Event Horizon: - Event Horizon is the outer edge of boundary of the black hole from which nothing can escape once crossed.

10. General Relativity: - The Theory of Gravitation developed by Elbert Einstein incorporating and extending the theory of special relativity, in other word general theory of relativity uses the concept

of space and time of special relativity. And General Theory of Relativity said that space and time will be relative but not absolute.

11. Geocentric Theory – Geocentric Theory said that, Earth were place at the center of the universe and all others stars, planet, are revolve around in a circular path.

12. Heliocentric Theory: - Heliocentric theory states that the sun place at the center of the universe and all other stars, planet revolve around in an elliptical orbit.

13. Light Year: - The distance travelled by a light in one year. therefore 1 light year $= 9.4607 \times 10^{12}$ Km

14. Neutron Star: - when a star is died in a supernova explosion a Neutron star will be formed . The size of Neutron star is typical about 11.75 km in radius. Neutron star is highly dense then White dwarf.

15. Parsec: - The distance to an object that has a parallax of one arc second which is equivalent to 3.26 light years.

16. Sagittarius A^* : - Sagittarius star A is a super massive Black Hole which is present in the core of Milky Way Galaxy.

17. Singularity: - Singularity point is present in the center of the black hole where all the matter of the black hole is present in only a single point.

18. Stadia: - Stadia is an unit of measurement of length given by Aristotle in old age . And Stadia are equal to 200 yards .But exactly not found the definition of stadia.

19. Supernova: - supernova is the large or highly explosion of the stars. When a star is die during a supernova explosion a white dwarf, Neutron stars, and black hole is formed.

20. White Dwarf:- An White Dwarf is formed when a star is collapse due to its own gravity .in other words when a star is died due to the cause of supernova explosion, then we get large amount of White Dwarf. However White Dwarf is less dense as compared to Neutron Star and Black Hole.

21. White Hole:-A white is in opposite to a Black Hole .We know that a Black Hole attract everything

that comes around it. And even Black Hole does not leave light .But in case of White Hole, it is a hypothetical region of space time, and it throws out everything that comes toward it.

www.ingramcontent.com/pod-product-compliance
Lightning Source LLC
Chambersburg PA
CBHW050452290526
45786CB00006B/2266